MICROSCOPES

Rebecca Woodbury, Ph.D., M.Ed.

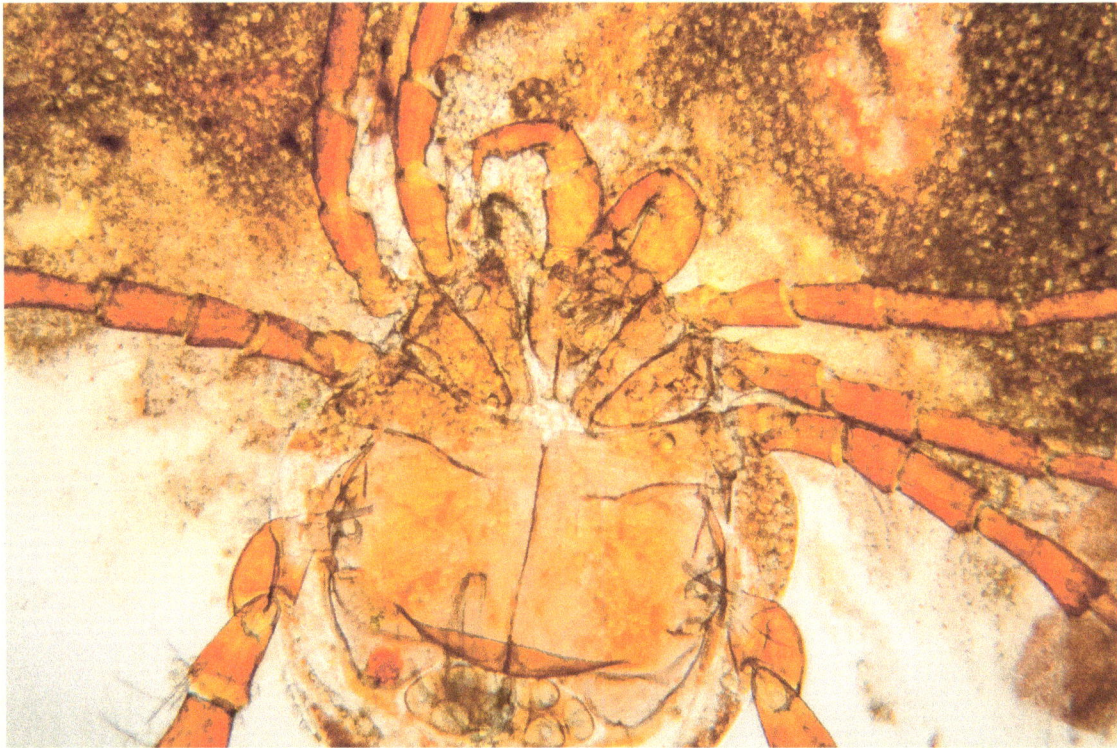

Gravitas Publications Inc.

Microscopes

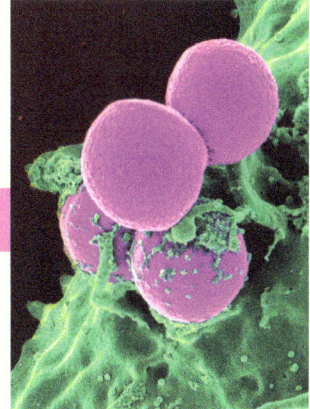

Illustrations: Janet Moneymaker

Microscopes
ISBN 978-1-950415-72-4

Published by Gravitas Publications Inc.
Imprint: Real Science-4-Kids
www.gravitaspublications.com
www.realscience4kids.com

RS4K

Image credits: Cover and Title Page, By ileana_bt, AdobeStock; Above, NIH Imagebank; P. 5. Blausen.com staff (2014). "Medical gallery of Blausen Medical 2014". WikiJournal of Medicine 1 (2). DOI-10.15347 wjm 2014.010. ISSN 2002-4436; P. 15. By Chatsikan, AdobeStock; P. 17. CDC, PaulHowell; P. 19. Rebecca Woodbury, Ph.D. Dissertation 1992; P. 20. – 1. By Viks_jin, AdobeStock; 2. By Frank Fox, www.mikro-foto.de, CC BY SA 3.0 Germany; 3. By ileana_bt, AdobeStock 4. Gregory Antipa (San Francisco State University) Public Domain; P. 21. – 1. NIH Imagebank; 2. MarkTalbot/CSIRO; 3. CDC–JaniceCarr/OrenMayer; 4. CDC/JaniceCarr; 5. NIAID

Can you see...

...the nose of a bug?

Can you see...

...a red blood cell?

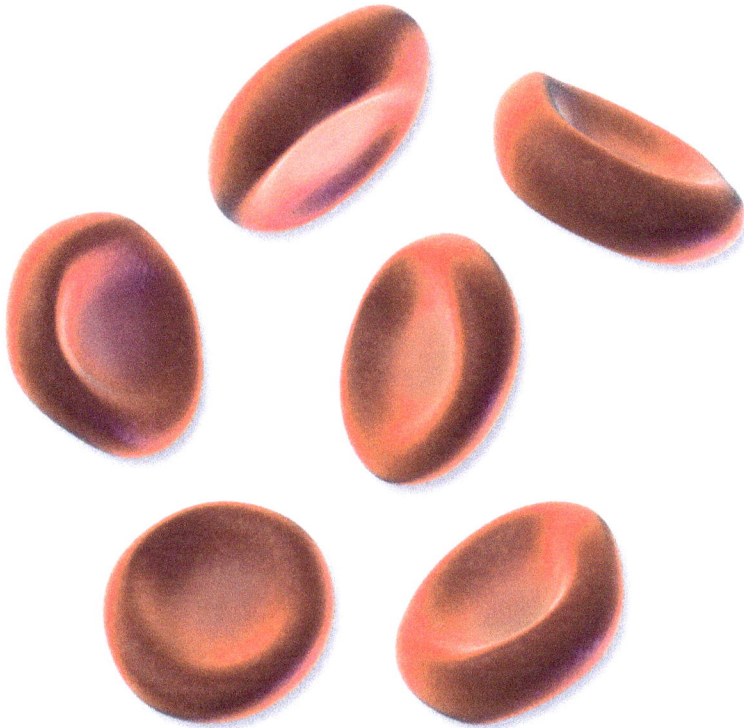

Can you see...

...an atom?

Some things are too small
to see with our eyes only.

I must be
BIG!

We can use a **microscope**

to see tiny things.

I want to see them too!

A **microscope** makes small things look bigger.

This is a **light microscope.**
A light microscope has a **lens**, a **stage**, and a **light source**.

What do the parts of a
light microscope do?

Lens: Magnifies the **sample.**
(Makes it look bigger.)

Stage: Holds the **sample**
(the object being
looked at).

Light: **Illuminates** the
sample.

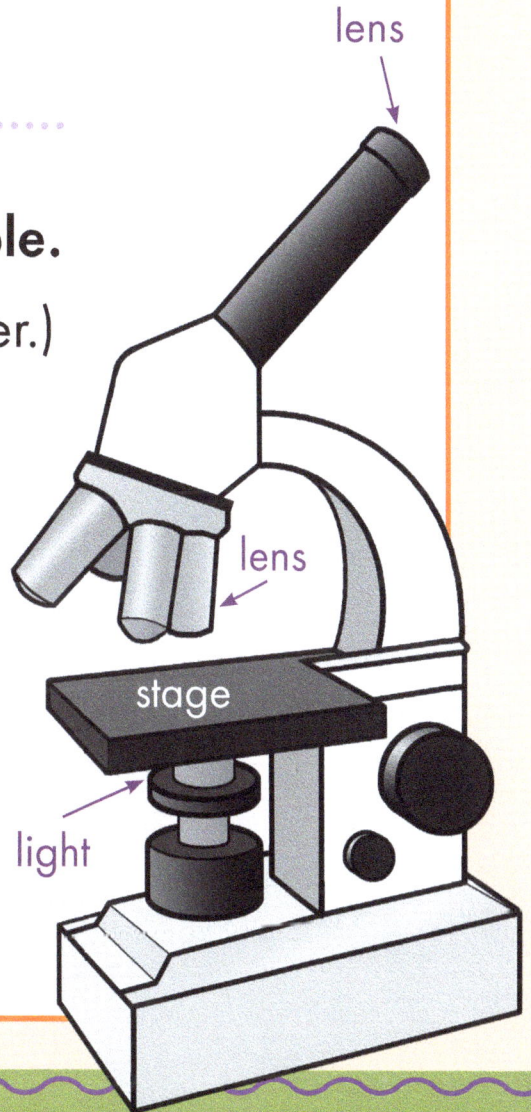

lens

lens

stage

light

With a light microscope

we can see red blood cells.

Do we have red blood cells?

Yes! We have blood.

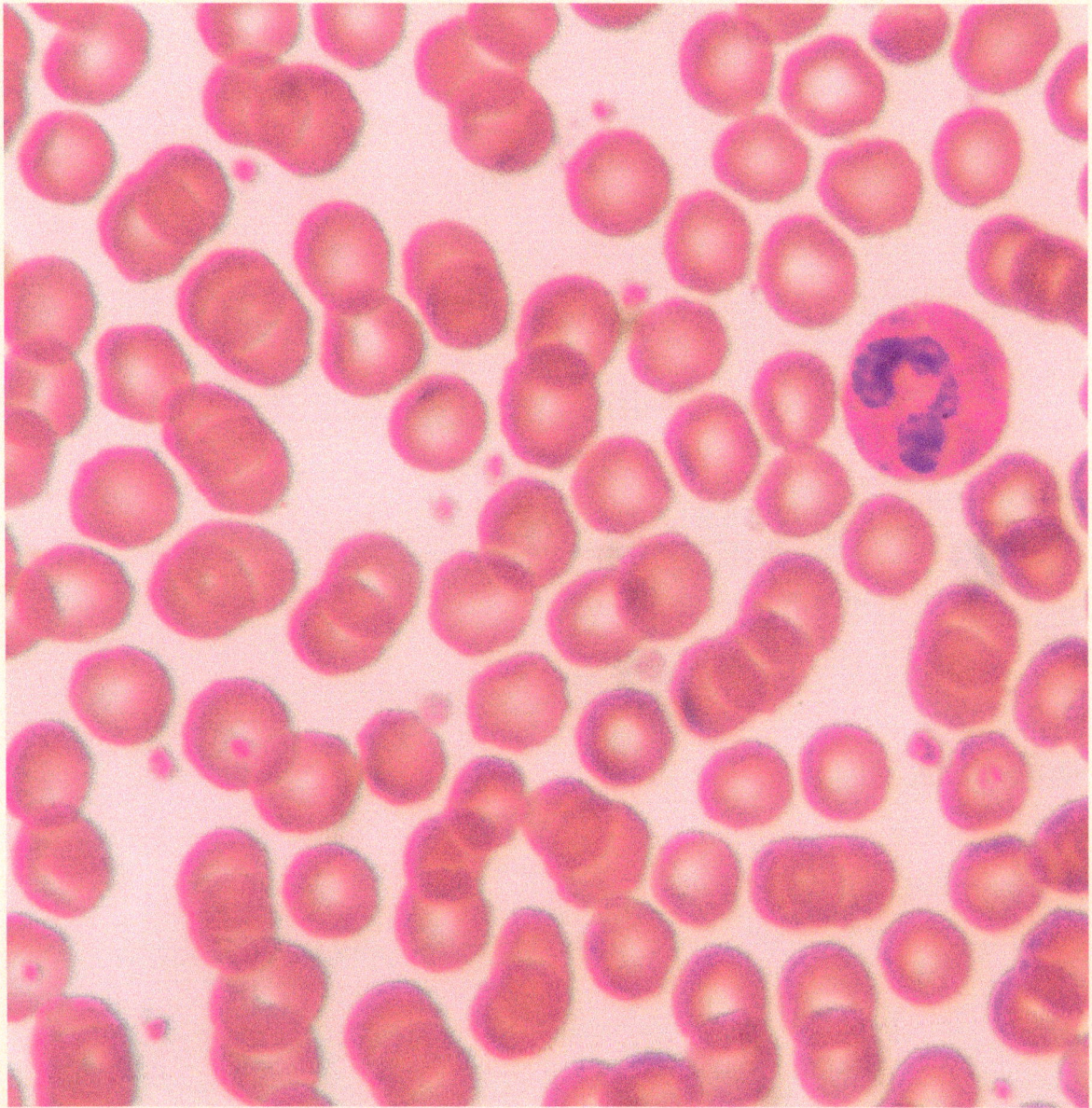

Another type of microscope

lets us see the nose of a bug.

I did not know bugs are so pretty!

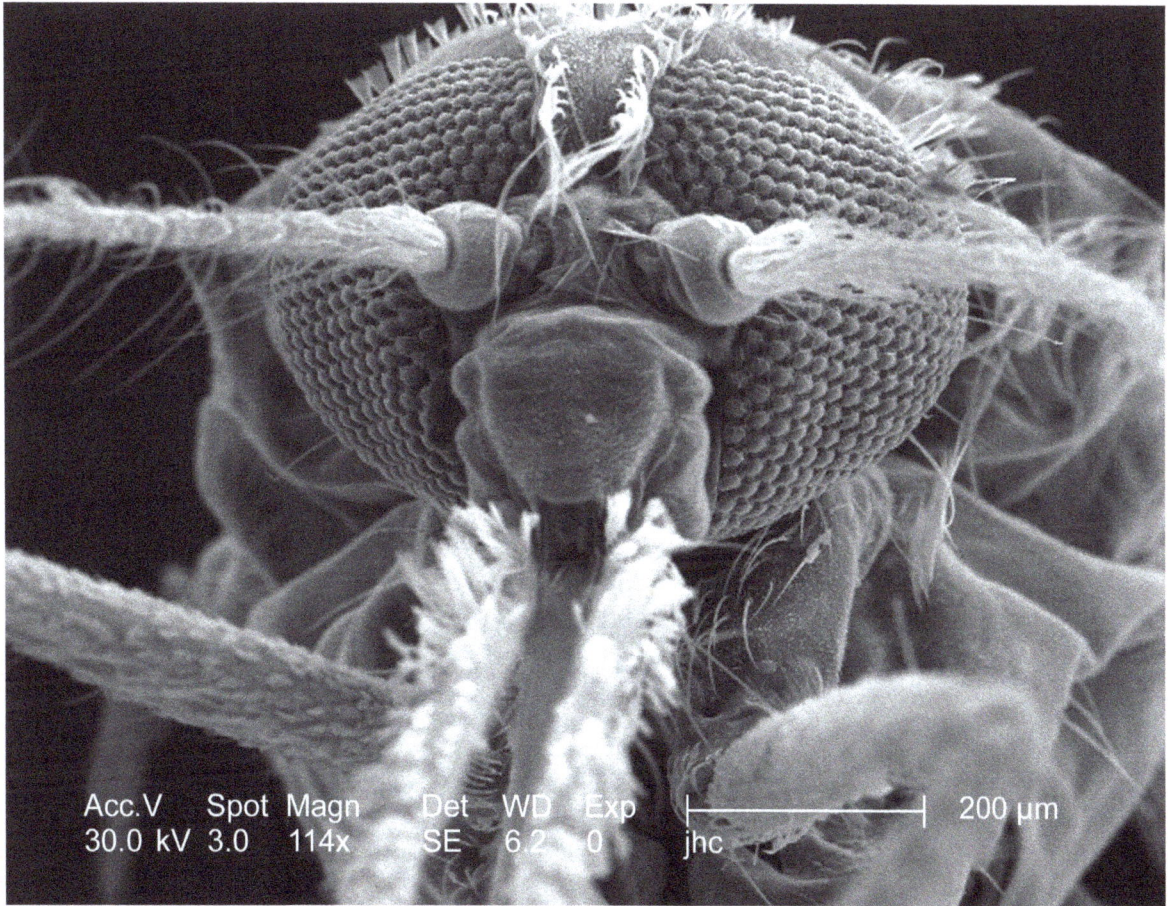

Mosquito Head

And yet another type of microscope lets us see atoms!

Atoms are pretty too!

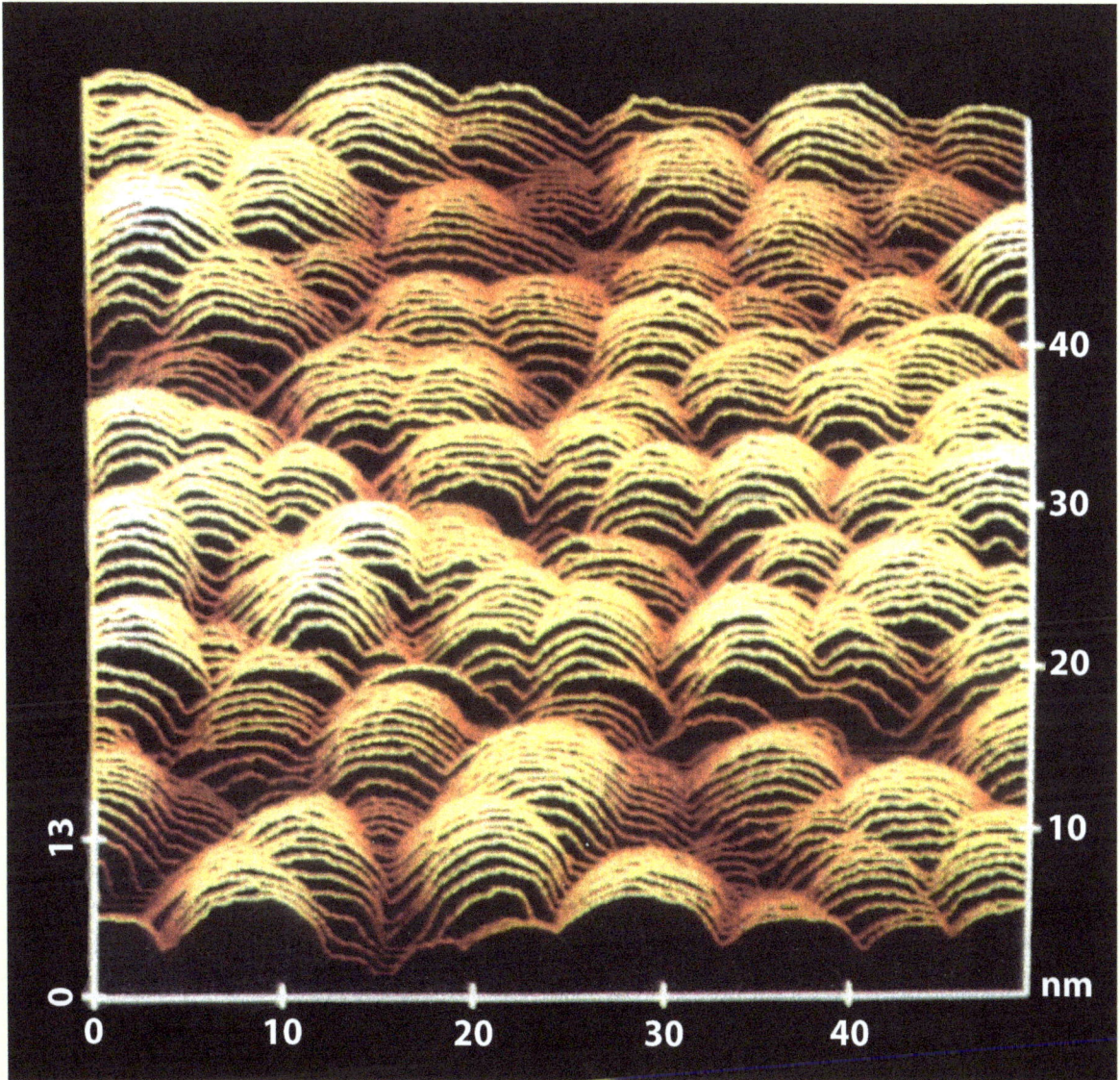

Carbon Atoms on a Surface

Microscopes help us see an entire tiny world!

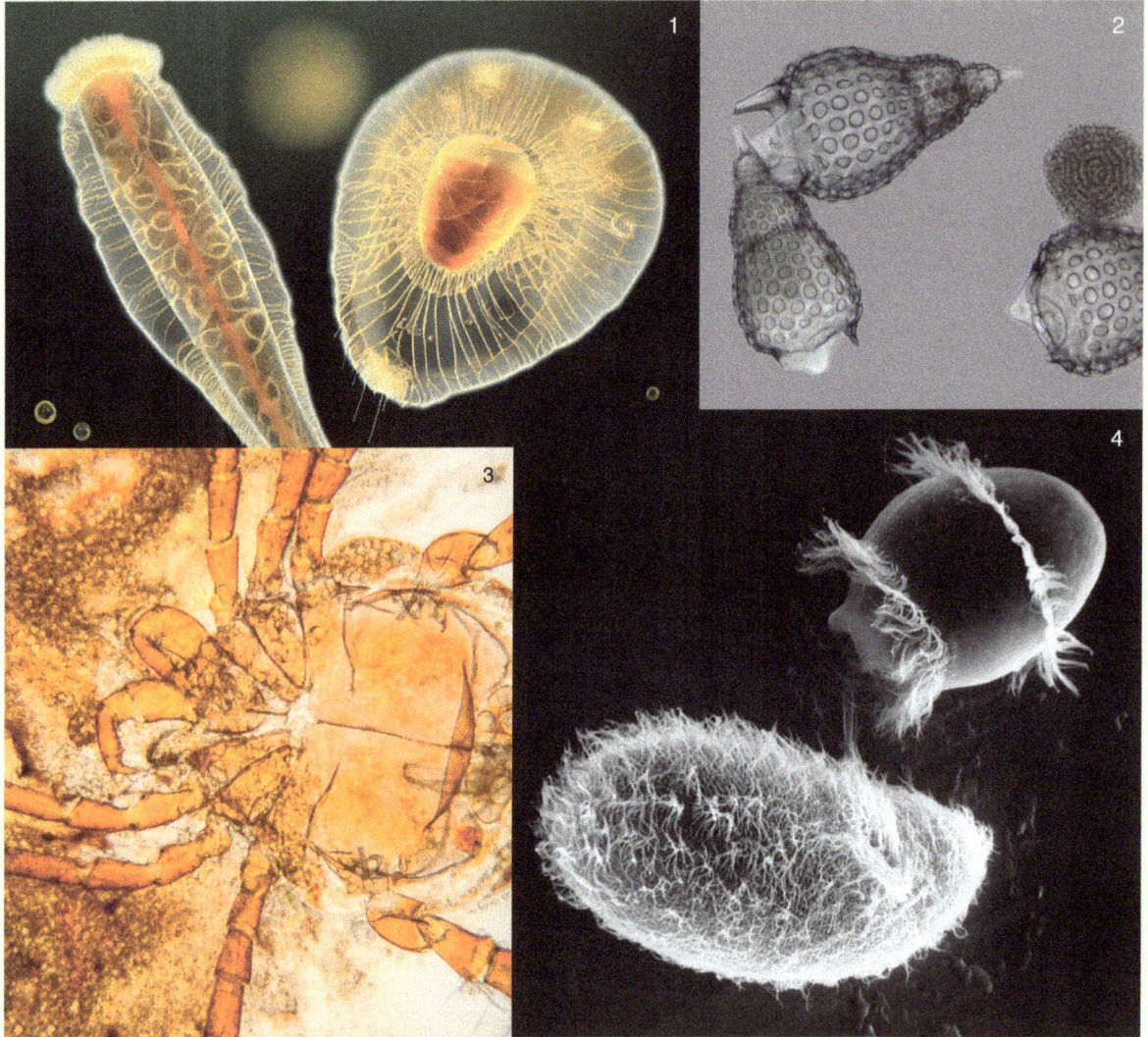

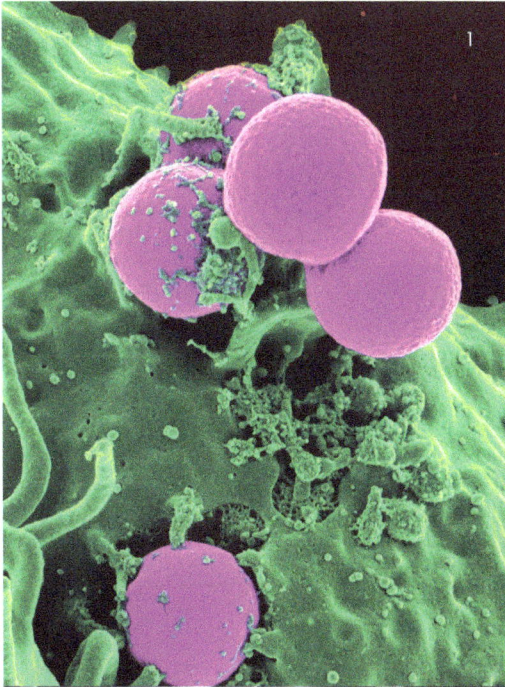

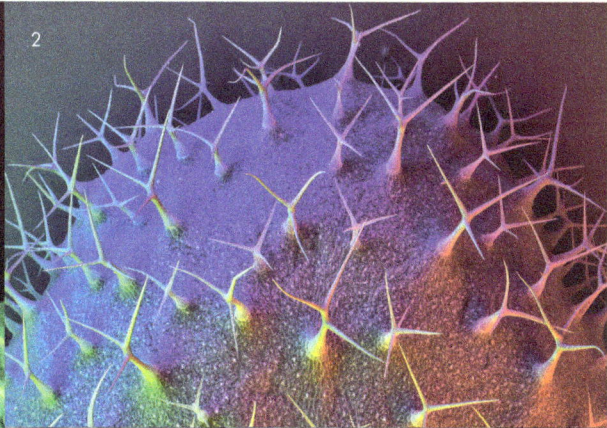

Acc V Spot Magn   Det WD Exp   2 µm
30.0 kV 3.0 11834x SE   7.2 0   jhc

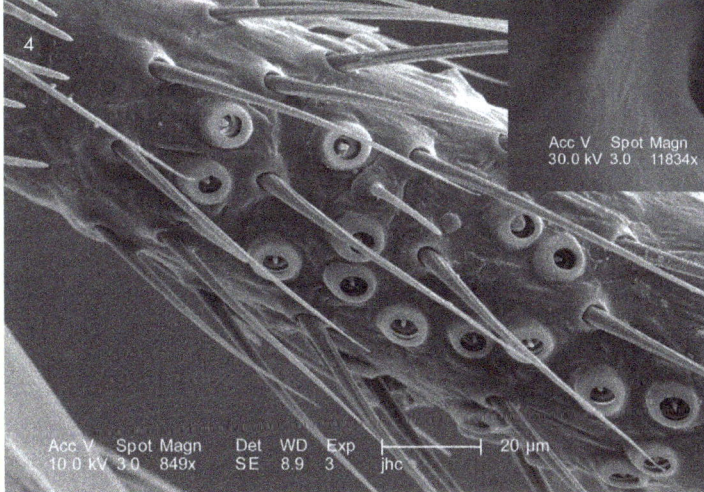

Acc V Spot Magn   Det WD Exp   20 µm
10.0 kV 3.0 849x   SE   8.9 3   jhc

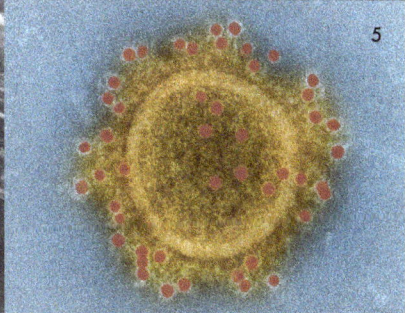

How to say science words

cell   (SEL)

illuminate   (ih-LOO-muh-nayt)

lens   (LENZ)

light source   (LIYT SAWRS)

magnify   (MAG-nuh-fiy)

microscope   (MIY-kruh-skohp)

microscopic   (miy-kruh-SKAH-pik)

sample   (SAM-puhl)

stage   (STAYJ)